Washington Augustus Roebling

Description of a new Method of transmitting Power by means of wire Ropes

Washington Augustus Roebling

Description of a new Method of transmitting Power by means of wire Ropes

ISBN/EAN: 9783337024581

Printed in Europe, USA, Canada, Australia, Japan

Cover: Foto ©berggeist007 / pixelio.de

More available books at **www.hansebooks.com**

DESCRIPTION

OF A

NEW METHOD

OF

TRANSMITTING POWER

BY MEANS OF

WIRE ROPES.

BY

W. A ROEBLING, C.E.,

TRENTON, N. J.

THIRD EDITION.

NEW YORK:

D. VAN NOSTRAND, PUBLISHER,

23 MURRAY STREET AND 27 WARREN STREET,

1872.

INTRODUCTORY REMARKS.

PART I.

" *The use of a round endless wire-rope running at a great velocity in a grooved sheave, in place of a flat belt running on a flat-faced pulley, constitutes the transmission of power by wire-ropes.*"

The distance to which this can be applied ranges from 50 or 60 feet up to about 3 miles. It commences at the point where a belt becomes too long to be used profitably, and can thence be extended almost indefinitely. In point of economy it costs only one-fifteenth of an equivalent amount of belting and the one twenty-fifth of shafting.

This method was first introduced, both in Europe and America, about the year 1850. The development it has received in this country is but trifling; in Europe, however, it has been immense, numbering at the present time over 2,000 permanent applications, and as many more of a temporary nature.

Visitors to the Paris exposition in 1867 doubtless remember seeing in the neighborhood of the iron lighthouse two slender ropes whizzing through the air high above their heads at the rate of a mile a minute, and passing in their course over the broad promenades, the garden and part of the artificial basin. Upon entering one of the buildings in which the ropes disappeared, they saw a huge centrifugal pump, raising a stream of water twelve inches in diameter from the reservoir below. In looking at the ceaseless flow, every one was impressed with the idea that the ropes formed some kind of a mysterious connection between this pump and a steam-engine working in a building three hundred feet off; few, however, understood it.

That was simply a *transmission* of *power by ropes.* The entire force of a 50-horse-power engine was thus conveyed through the air by one endless half-inch rope, and was consumed in driving "M. Schneider's great pump."

There is scarcely an establishment in existence where it would not be convenient at times to transfer power to some isolated building located at a distance, without going to the trouble or expense of putting up an engine.

Here we have the ready means at command—a means which recommends itself by its cheapness of first cost, its economy of maintenance, and perfect reliability in regard to working. To enumerate all the instances where it can be applied would be too formidable a task,—a few, however, will be of interest, and will readily lead the reader to fill out the list for himself.

Many valuable sites for water-power are lying idle in this country for want of building-room in their immediate vicinity; and since the water can only be led down hill in certain directions, the cost of a canal or flume would in many cases come too high, and so the power remains unimproved. By ropes, however, we can convey the power of a turbine or waterwheel in any direction, both up stream as well as down stream, to either side if necessary; up an ascent of 1 in 8 or 10, or down a moderate slope as well. The power need not be confined to one factory, but can be distributed among a dozen, located so as to suit their particular business, and not to suit the oftentimes inconvenient location of a canal. If the water-power is on one side of the river and the factories on the other, it is an easy matter to transfer it across, by making one or two artificial stations in the river, which nature often supplies by a rock in place. *(See Frontispiece.)*

In the neighborhood of Frankfort-on-the-Main, in Germany, the power of a 100-horse-power turbine is conveyed for a distance of 3,200 feet, by means of a rope-transmission, to a cotton factory located in the proper place for such a building. Wheels of 13½ feet, making 114 revolutions per minute, are used ; size of rope ⅝ inch, stations 8 in number and 400 feet apart. A nearer site for a building could not be found, and this was the only way in which the power could be made available for that purpose.

At another establishment—a powder-mill—the various buildings were placed 400 feet apart along the circumference of a circle having a diameter of about 1,200 feet. In the centre a waterwheel supplied the power, which was conveyed to each building by a rope-transmission. One man at the central building sets in motion the machinery of all the buildings, which on this account could be placed far enough apart to prevent the explosion of one from passing to the other.

In many factories, long counter-shafting with heavy bevel-gearing can be saved by using a rope ; the farther off the shaft we wish to drive, the better the arrangement will work.

A heavy punch or pair of shears, straightening-rolls, etc., may at times be more conveniently located out in the yard, near the metal which they are to work upon ; but in the ordinary way it would be rather troublesome to convey power to them, and so they are put up inside, and the metal is carried in and out at a heavy annual expense, all of which could be saved by this method.

It can be profitably applied as a substitute for horse-power used at outdoor work by rolling-mills, furnaces, mines, and all sorts of contractors' building operations.

Factories in cities are generally cramped for room. When neighboring property cannot be bought, perhaps that across the street can : yet the trouble and expense of digging up the street to lay down a line of shafting is sufficient to deter one from the purchase. For such a case a remedy is here presented. A little endless rope passing through a couple of slits in the window-casing of an upper story, across to the story opposite, will do all the work, and none of the passers-by will be any the wiser for it. A belt would require protection from the weather, but the rope does not, and can hang free in the air.

From an engine in the basement, power can be readily conveyed to the upper stories; it is necessary, however, that for a certain distance the rope should hang horizontally, in order to gain the required tension.

As the largest example of a wire-rope transmission we may mention the great improvement at the Falls of the Rhine, near Schaffhausen, in Switzerland: advantage was taken of the rapids at one side and a number of turbines put in, aggregating in all 600 horse-power. Since the steep rocky banks forbade the erection of any factories in the immediate neighborhood, the entire power was transferred diagonally across the stream to the town, about a mile lower down, and there distributed, certain rocks in the water being made use of to set up the required intermediate stations. There are, no doubt, hundreds of similar localities in our country which can be improved in this way.

New England especially abounds with them. Coal being so expensive there, their value is all the greater. At the same time the rough and rocky nature of most of her river banks has in many cases proved a barrier to the erection of factories near by. Now, however, by the system of rope-transmission we can devote all this waste power to a useful purpose.

For much of the material embraced in the following pages I am indebted to the Swiss brothers Hirn, who have been mainly instrumental in developing the system practically on the continent; and also to Prof. F. Reuleaux, who treats of the subject in his "Constructeur."

Diam. of Wheel in Feet.	No. of Rev.	Trade No. of Rope.	Diam. of Rope.	Horse Power.	Diam. of Wheel in Feet.	No. of Rev.	Trade No. of Rope.	Diam. of Rope.	Horse Power.
4	80	24	3/8	3.3	10	80	19/18	5/8 11/16	55. / 58.4
4	100	24	3/8	4.1	10	100	19/18	5/8 11/16	68.7 / 73.
4	120	24	3/8	5.	10	120	19/18	5/8 11/16	82.5 / 87.6
4	140	24	3/8	5.8	10	140	19/18	5/8 11/16	96.2 / 102.2
5	80	23	7/16	6.9	11	80	19/18	5/8 11/16	64.9 / 75.5
5	100	23	7/16	8.6	11	100	19/18	5/8 11/16	81.1 / 94.4
5	120	23	7/16	10.3	11	120	19/18	5/8 11/16	97.3 / 113.3
5	140	23	7/16	12.1	11	140	19/18	5/8 11/16	113.6 / 132.1
6	80	22	15/32	10.7	12	80	18/17	11/16 3/4	93.4 / 99.3
6	100	22	15/32	13.4	12	100	18/17	11/16 3/4	116.7 / 124.1
6	120	22	15/32	16.1	12	120	18/17	11/16 3/4	140.1 / 148.9
6	140	22	15/32	18.7	12	140	18/17	11/16 3/4	163.5 / 173.7
7	80	22	15/32	16.9	13	80	18/17	11/16 3/4	112. / 122.6
7	100	22	15/32	21.1	13	100	18/17	11/16 3/4	140. / 153.2
7	120	22	15/32	25.3	13	120	18/17	11/16 3/4	168. / 183.9
7	140	22	15/32	29.6	14	80	17/16	3/4 7/8	148. / 141.
8	80	21	1/2	22.	14	100	17/16	3/4 7/8	185. / 176.
8	100	21	1/2	27.5	14	120	17/16	3/4 7/8	222. / 211.
8	120	21	1/2	33.	15	80	17/16	3/4 7/8	217. / 217.
8	140	21	1/2	38.5	15	100	17/16	3/4 7/8	259. / 259.
9	80	20/19	11/32 5/8	40. / 41.5	15	120	17/16	3/4 7/8	300. / 300.
9	100	20/19	11/32 5/8	50. / 51.9					
9	120	20/19	11/32 5/8	60. / 62.2					
9	140	20/19	11/32 5/8	70. / 72.6					

PART II.

On the preceding page a table of horse-powers is presented. It embraces every case that will ordinarily arise in practice, and one can readily select that combination which will suit his own case, especially if the driving machinery already exists.

The first column gives the diameters of the grooved sheave-wheels, in which the rope runs, commencing with 4 feet. Smaller wheels are but seldom wanted.

Then knowing the number of revolutions which your shaft makes, the last column gives the horse-power corresponding to a certain-sized wheel.

Where there is a choice between a small wheel and fast speed, or a larger wheel with slower speed, it is recommended to take the larger wheel.

The horse-powers here given are a minimum, and can be relied upon under all circumstances.

The Driving Ropes.

The range in the size of wire-ropes is small, varying only from ⅜ in. to ¾ in. diameter in a range of 3 to 250 horse-power. Full information concerning the strength, cost, etc., of the ropes is contained in the "Wire-rope table" on the last page of this pamphlet. The ropes are always kept on hand, and can be spliced endless at the factory ; or else a man is sent to splice them, whenever an endless rope cannot be put on direct. Two wire-rope transmissions can also be seen in operation at the factory.

In regard to cost, they are the cheapest part of a transmission. For instance, a No. 22 rope, conveying say 25 horse-power costs 8 cents per foot, whereas an equivalent belt costs about $1.40 per foot. Where a rope-transmission has to be constantly at work, it is good policy to keep a spare rope on hand ready spliced, so as to avoid delay. Their duration is from 2½ to 5 years, according to the speed.

For the smaller powers it is advisable to take a size larger, for the sake of getting wear out of the rope; although it must be borne in mind that a larger rope is always stiffer than a small one, and therefore additional power is lost in bending it around the sheave. An illustration of that is seen in the case of the 14-feet wheel in the table, where a ⅞-rope gives less power than a ¾-rope, simply because it is so much stiffer.

Ropes for this purpose are always made with a hemp core, to increase their pliability.

Equivalent Belt.

It is often required to convey the entire power of a certain shaft which is driven by a belt of a given size. In such a case, a simple rule agreeing with the average result of practice is, that 70 square feet of belt-surface are equal to one horse-power.

Take, for example, a belt 1 foot wide running at the rate of 1,400 feet per minute; then the

$$\text{Horse-power} = \frac{1400' + 1'}{70'} = 20;$$

and by referring to the table we find the diameter of the wheel corresponding to this horse-power, and making the same number of revolutions that the belt-pulley does.

Distance of Transmission.

The foregoing table is intended for distances from 80 up to 350 or 400 feet in one stretch. For a single stretch extended to say 450 feet, where no opportunity is presented for putting in an intermediate station, we must use a rope one size heavier ; and in a case where there is not sufficient head-room to allow the rope its proper sag, and it has to be stretched tighter in consequence, we must also take a rope one size heavier.

Short Transmission.—Whenever the distance is less than 80 feet, the rope has to be stretched *very tight*, and we no longer depend upon the sag to give it the requisite amount of tension. Here we must take a rope two sizes heavier than is given in the table, and run at the maximum speed indicated : it is also preferable to substitute in place of the rope of 49 wires, a fine rope of 133 wires of the same diameter, which possesses double the flexibility, runs smoother, and lasts longer. In fact, the substitution of a fine rope for a coarse one can be done with advantage in every case in the table where the size admits of it.

Splices.—Both kinds of rope are spliced with equal facility. The splices are all of the kind known as the long-splice ; the rope is not weakened thereby, neither is its size increased any, and only a well practised eye can detect the locality of one.

Relative Height of Wheels.—It is not necessary that the two wheels should be at the same level, one may be higher or lower than the other without detriment; and unless this change of level is carried to excess, there need be no change in the size of wheel or speed of rope : the rope may have to be strained a little tighter. As the inclination from one wheel to another approaches an angle of 45°, a different arrangement must be made, as will be shown hereafter on page 21.

Deflection or Sag of the Ropes.

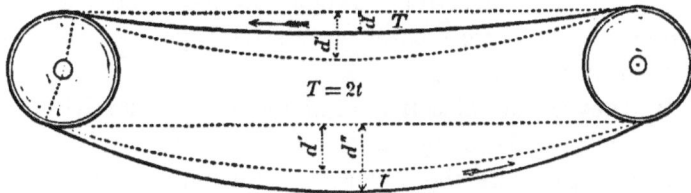

In the above illustration the upper rope is the pulling-rope and the lower one the loose following-rope. When the rope is working, the tension T in the upper rope is just double that in the lower rope, hence the latter will sag much lower below a horizontal line than the upper one.

When the rope is at rest, both ropes will occupy the position indicated by the dotted line, and will have a uniform tension.

The best way in practice is to hang up a wire in the position the rope is to occupy at rest: that has to be done in any case, in order to get the length of rope needed. Then hang it so that the deflection d', below the horizontal line, is about $\frac{1}{15}$th of the whole distance from wheel to wheel. The deflection d of the upper running-rope will then be about $\frac{1}{15}$th to $\frac{1}{20}$th.

The deflection d'' of the lower working-rope is on an average one-half greater than the deflection d' of the rope at rest. This is of importance, as we should know beforehand whether the lower rope is going to scrape on the ground or. touch other obstructions; in that case, we either have to dig a trench for the lower rope to run in or else raise both wheels high enough to clear.

Practically, however, it is not necessary to be so particular about this matter, on account of the stretch in the rope. Wire-rope stretches comparatively very little;

still there is some stretch, and it is well to allow for it by stretching the rope a little too tight at first; after running a little it will hang all right. When the rope is very long it is advisable to take up the stretch at the end of two or three months, as a slack rope does not run so steadily as a tight one.

Whenever the direction of the motion of the driving wheel is not fixed by other circumstances, it is often advisable to make the lower rope the pulling-rope, and the upper the follower, as here shown. In this way obstructions can be avoided, which by the other plan would have to be removed. The ropes will not interfere as long as the difference between the two deflections d' and d'' is less than the diameter of the wheel.

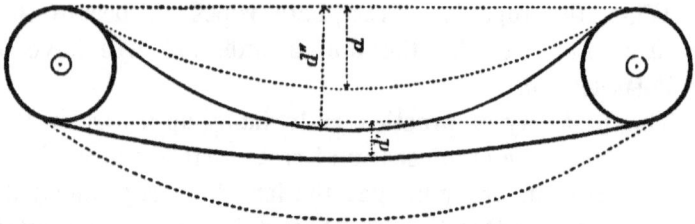

These limits are of use whenever, on account of rocks or otherwise, we have to move the wheels closer together, and the question is how far to have them apart with a certain deflection.

The Wheels

are generally made of cast-iron, with a stout hub, 8 curved arms, and a deep, flaring groove.

: On account of the great centrifugal force of a rapidly

revolving wheel, a wooden rim would not answer. The section here shown is for a single grooved wheel. The slope of the sides of the rim should be considerable ; it has been made as high as 45° in some instances, where

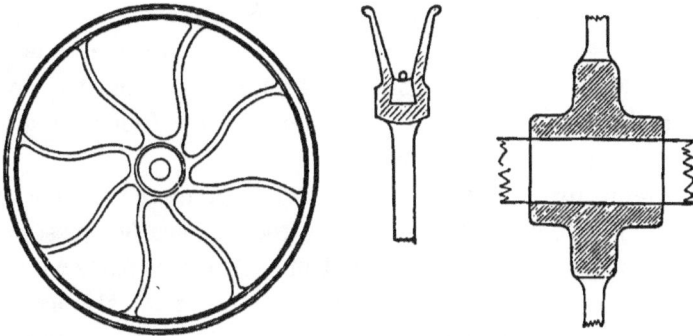

the span was very long and the ropes were exposed to a high side-wind. But the half of this slope will answer in general.

"*A set of patterns of these wheels, single-grooved, from 4 feet to 12 feet diameter, is kept on hand at the wire-rope works in Trenton, N. J., and castings can be furnished at short notice.*"*

The bottom of the groove is made a little wider, to prevent the filling from flying out. The rope should always run on a cushion of some kind, and not on the iron, which quickly wears it out. A variety of material is used for this filling—soft wood, india-rubber, leather, old rope tarred, and oakum. To use end wood the rim has to be constructed on a different plan from that shown here. The objections to it are, that it is liable to shrink and crack and fly out ; it is also more severe on the rope. India-rubber is a very good material ; strips of an inch square or less can be wedged in very quickly, and will last a long time. We use it now exclusively for the filling of our wheels.

---

* See page 22 for price of wheels.

The rubber is cut into short pieces, having a cross-section, as here shown, and is made larger than the groove, so that when once forced in, it cannot fly out. The adhesion of the rope is likewise greater on the rubber than on any other material.

Leather has been used to some extent. It is durable, but tedious to put in, as the thin strips of leather must be set in on end, and several thousand are required for a large wheel.

Again, by wedging the groove full of tarred oakum a filling is also obtained, nearly as good as leather, costing less, and not so tedious to put in. Another plan, which I have tried with success, is to revolve the wheels slowly and let a lot of small-sized tarred ratlin or jute-yarns wind up on themselves in the groove ; then secure the end, and after a day or two of running the pressure of the rope, together with the tar, will have made the filling compact. This makes a cheap filling.

The double-grooved wheels are filled in the same way.

The rope will run on such filling without making any noise whatever, and soon wears in a round groove for itself.

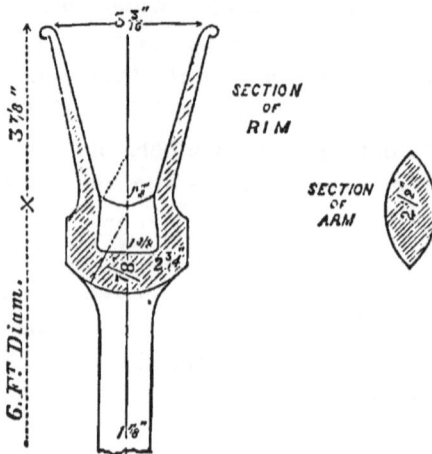

A section of the rim of a 6-foot wheel is here shown with the dimensions marked.

The diam. of the wheel is not reckoned from the outside of the rim, but from the top of the filling, which corresponds to the circle

PLATE II.

described by the rope. The hub is made of ample size, so as to admit of being bored out for shafts, varying from 2 to 3¼ inches.

"Special care must be taken to set the wheel-shaft at right angles to the line of the transmission, and also to set the wheel square with the shaft, otherwise the wheel will wabble, and cause the rope to vibrate and jerk."

In conveying power from one building to another at a single stretch, it is often most convenient to extend the driving-shafts through the wall and have the wheels and rope running free outside. See Plate II. The endless spliced rope can be laid on directly in this case, which is often an advantage. When this is not practicable, and the rope has to run through the wall or the side of the window-casing, narrow slits should be cut in, from 9 to 12 inches high : these slits at the same time serve as guides to lead the rope to the centre of the wheel-groove. Another variation would be, to set up the wheels on the roof, where they are entirely out of the way. The rope while running requires no protection. If it has to stand still much, pour some hot coal-tar from a can on the rope in the groove of the wheel while running.

Whenever there is no room for the sag of the rope, and it is inconvenient to raise the wheels higher, or a ditch cannot be dug, it may be supported by a roller in the

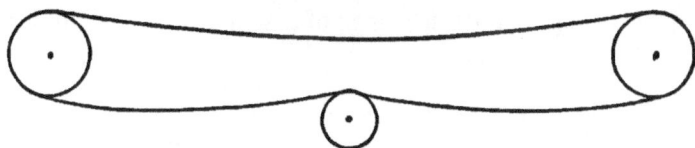

middle. This supporting-roller must be in the centre of the span, and must be at least half the size of the larger wheels.

LONG TRANSMISSIONS.

SEE PLATE III.

———— ◆ ————

WHEN the distance materially exceeds 350 to 400 feet, a rope-transmission should be divided into 2 or more *equal* parts, by means of one or more intermediate stations. At each station there is a wheel mounted on a pedestal or other support, and provided with a double groove in the rim ; so that in place of one long continuous rope, we have 2 or more shorter endless ropes, extending from station to station. This is far preferable to supporting-rollers in the middle, especially when the demand on the power is intermittent and jerks would thereby be caused in the rope. With the two-grooved wheel that cannot take place : moreover, the wear of the rope on a supporting-pulley is greater. The sketch on the adjoining page gives a view of the arrangement.

The whole system should be in a straight line from end to end. The number of stations can be extended indefinitely.

Transmissions are in operation a mile in length. The loss of power from friction, etc., or bending of rope, does not amount to 10 per cent. per mile, and need not be taken into account at all for only one station.. No slipping of the rope in the groove ever occurs with a proper filling. With bearings of a sufficient length under the shaft of the centre wheel, and by providing them with a self-feeding oil-cup, the axle-friction is reduced to a minimum.

Compare this now with a line of shafting where a bearing has to be provided every 12 or 15 feet, whereas here we need a bearing only every 3 or 400 feet. Shafting is simply out of the question in such a case.

The cuts on pages 17 and 19 present three varieties of foundation or pedestal for the two-grooved wheel—one of stone, one of cast-iron, and the last of wood.

In this country it will generally be found cheaper to put up a wooden frame, bolted to a masonry foundation extending below the reach of frost. The frame should be braced from each side so as to maintain the wheel in a vertical plane: end-bracing is not required. The length of shaft from centre to centre of bearing should be a little

less than half the diameter of the wheel. A collar must be put on the shaft on the inside of each bearing.

It is not necessary, however, that the wheel should be set in the middle between a double frame or pedestal ; we can just as well hang it free on the outside, as indicated in the dotted lines of the outer wheel—Fig. 2, previous page. The great advantage this latter arrangement gives us, is, that we can mount a rope ready spliced, simply by laying it on from the side. In the other way, the rope has to be rove around, hauled taut by a fall, and the splice made at the spot.

In a short transmission it is generally more convenient to put on the rope ready spliced ; but where there is a large number of stations and many ropes, the man in charge of them must learn to make his own splices. This is an easy operation, and can be learned in a few hours by anybody. The taking off of an old rope, and putting on the new one, including the splicing, should not take more than 1 or 2 hours.

This is of importance when the whole motive-power of a factory is derived from ropes.

Where ropes have to be often mounted it is convenient to use a short curved trough of angle-iron, first applied by Mr. Ziegler. This crosses the main groove at one end, and is secured to the rim and arms of the larger wheel by bolts or lashing. Upon turning the wheel in the direction of the arrow, the rope lays in itself. We can also ease the strain on the rope by putting under a light temporary support in the middle.

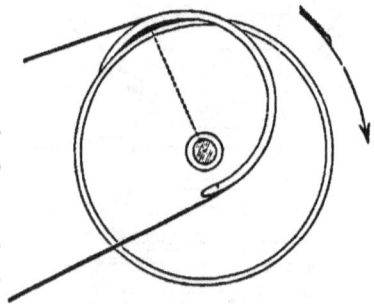

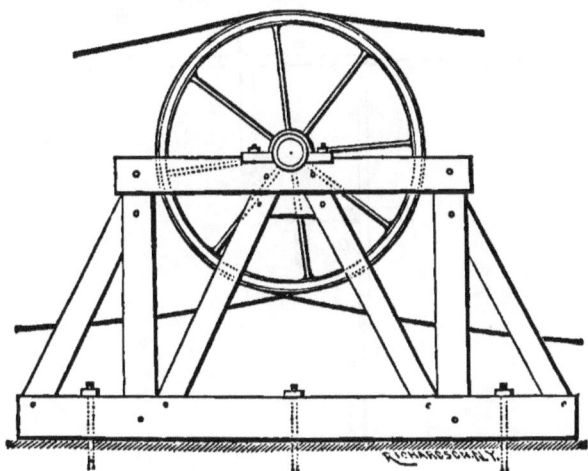

Plate III presents an illustration of a Turbine wheel
with the driving-wheel above, on the outside of the building.
This can be doubled by having a driving-wheel on the opposite side also, from which the ropes can pass in a different direction if necessary. The sketch is a little out of
proportion, as the distances between stations ought to be
three-fold, according to a scale.

It is required sometimes to change the direction of the
transmission at some point in its course, either to avoid
an obstacle, or for the purpose of distributing the power
to a number of consumers. This could be done by means
of horizontal sheaves, but the best method has been shown
by experience to consist in the use of bevel-wheels, as
shown in this sketch.

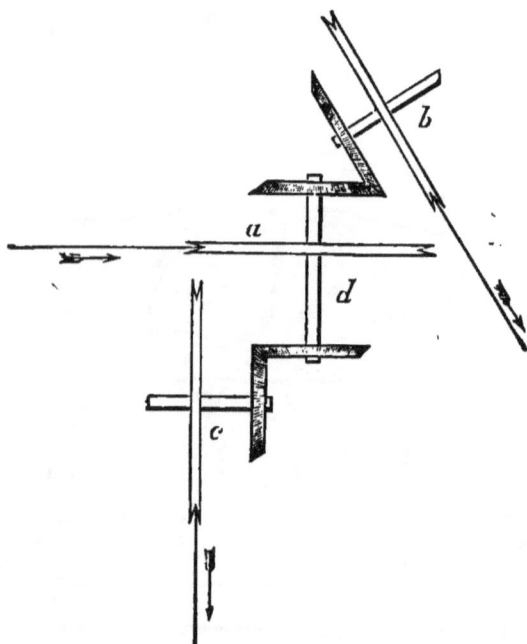

PLATE III.

This is called a distributing-station; *a* is the main driving-wheel; the two wheels *b* and *c* convey the power in the direction of the arrows.

We can thus reach any locality desired, or get around the corners of buildings, or any part of your neighbor's property without difficulty. By lengthening out the shaft *d* in one direction, we can branch out still further.

When the power is to be conveyed nearly vertically, no good result is obtained by running the rope, say from *a* to *b* direct, as indicated by dotted lines in the figure below, since it would slip. Two carrying-sheaves, *c* and *d*, must be put up vertically above *a*, giving a horizontal stretch from *c* and *d* to *b*. This is necessary, in order to maintain the required tension in the rope, which can be obtained in no other way. *a*, *b*, and *c*, and even *d*, should be of the same size; yet *d*, which supports the following-rope, may be made smaller without damage.

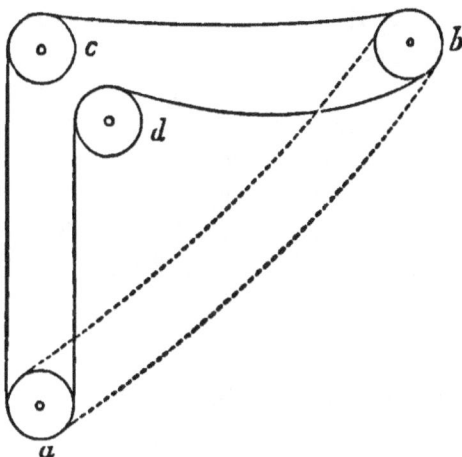

This arrangement must be borne in mind whenever the source of power is located in the cellar, and we

want to carry it to an upper story and distribute it horizontally.

In buildings we are often so cramped for room that
pulleys beyond 18 inches cannot be applied ; these would
be rather severe on a wire-rope, running fast, and a hide-
rope might be preferable.

On the last page will be found a Wire-rope table,
giving the particulars of the ropes called for by the
Horse-power table on page 7.

PRICE OF WHEELS,

FILLED WITH RUBBER AND BORED OUT.

3 feet diam.....	$30 each.	25
4 "	40 "	33
5 "	50 "	53
6 "	70 "	75
7 "	95 "	
8 "	125 "	
9 " CAST IN HALVES.	225 "	
10 " do. do.	300 "	
11 " do. do.	350 "	
12 " do. do.	400 "	
13 "		
14 "		
15 "		

Special rates for large wheels.

TABLE OF WIRE ROPE,

MANUFACTURED BY

JOHN A. ROEBLING'S SONS,

TRENTON, N. J.

ROPE OF 133 WIRES.

Trade Number.	Circumference in inches.	Diameter.	Price per Foot, in cts.	Ultimate strength in tons of 2,000 lbs.	Circumference of Hemp Rope of equivalent strength in inches.
1	6¾	2¼	1 20	74 00	15½
2	6	2	1 05	65 00	14½
3	5½	1¾	91	54 00	13
4	5	1⅝	78	43 60	12
5	4⅜	1½	65	35 00	10¾
6	4	1¼	53	27 20	9½
7	3½	1⅛	41	20 20	8
8	3⅛	1	34	16 00	[illegible]
9	2¾	⅞	28	11 40	6
10	2¼	¾	25	8 64	5
10¼	2	⅝	24	5 13	4½
10½	1⅝	9/16	23	4 27	4
10¾	1½	½	22	3 48	3¾

Tiller Rope, ⅜ in. diam., 26 cts.

Ropes from No. 8 to No. 10¾ are specially adapted for hoisting-rope.

Steel Rope of all sizes made to order.

All kinds of shackles, sockets, swivel hooks, and fastenings put on, and splices made.

ROPE OF 49 WIRES.

Trade Number.	Circumference in inches.	Price per Foot, in cts.	Ultimate strength in tons of 2,000 lbs.	Circumference of Hemp Rope of equivalent strength in inches.
11	4⅝	54	36 00	10¾
12	4¼	47	30 00	10
13	3¾	41	25 00	9½
14	3⅜	35	20 00	8¼
15	3	29	16 00	7¼
16	2⅝	23	12 30	6¼
17	2⅜	18	8 80	5½
18	2⅛	15	7 60	5
19	1⅞	13	5 80	4¾
20	1⅝	11	4 09	4
21	1⅜	9	2 83	3¼
22	1¼	8	2 13	2¾
23	1⅛	7	1 65	2½
24	1	6½	1 38	2¼
25	⅞	6	1 03	2
26	¾	5½	0 81	1¾
27	⅝	5	0 56	1½
27½	½	4		
28		3	Large Sash Cord	
29		2	Small " "	

Copper Rope corresponding to the above sizes, made to order.

Directions for Making a Long Splice in an Endless Running Rope, of Half Inch Diameter.

PLATE IV.

Tools required : One pair of nippers, for cutting off ends of strands; a pair of pliers, to pull through and straighten ends of strands ; a point, to open strands ; a knife, for' cutting the core, and two rope nippers, with sticks to untwist the rope ; also, a wooden mallet.

First.—Haul the two ends taut, with block and fall, until they overlap each other about 20 feet. Next, open the strands of both ends of the rope for a distance of 10 feet each; cut off both hemp cores as closely as possible (see Fig. 1), and then bring the open bunches of strands face to face, so that the opposite strands interlock regularly with each other.

Secondly.—Unlay any strand, *a*, and follow up with the strand 1 of the other end, laying it tightly into the open groove left upon unwinding *a*, and making the twist of the strand agree exactly with the lay of the open groove, until all but six inches of 1 are laid in, and *a* has become 20 feet long. Next, cut off *a* within six inches of the rope (see Fig. 2), leaving two short ends, which must be tied temporarily.

Thirdly.—Unlay a strand, 4, of the opposite end, and follow up with a strand, *f*, laying it into the open groove, as before, and treating it precisely as in the first case (see Fig. 3). Next, pursue the same course with *b* and 2, stopping, however, within four feet of the first set ; next, with *e* and 5 ; also, with *c*, 3, and *d*, 4. We now have the strands all laid into each other's places, with the respective ends passing each other at points 4 feet apart, as shown in Fig. 4.

Plate IV

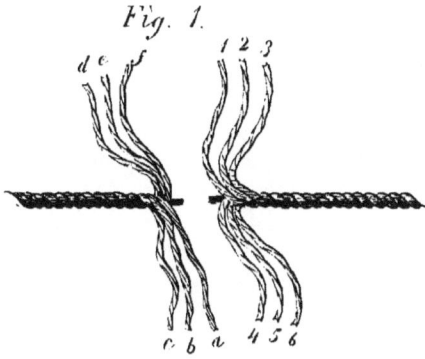

Fig. 1.

Fig. 4.

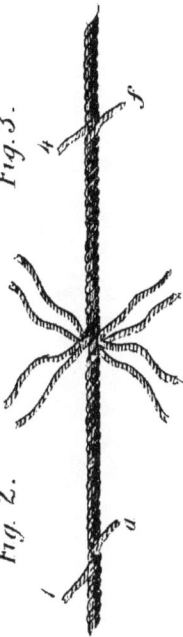

Fig. 2.

Fig. 3.

Fig. 5.

Fourthly.—These ends must now be secured and disposed of, without increasing the diameter of the rope, in the following manner : Nipper two rope slings around the wire rope, say six inches on each side of the crossing point of two strands. Insert a stick through the loop, and twist them in opposite directions, thus opening the lay of the rope (see Fig. 5). Now, cut out the core for six inches on the left, and stick the end of 1 under *a*, into the place occupied by the core. Next, cut out the core in the same way on the right, and stick in the end of *a* in place of the core. The ends of the strands must be straightened before they are stuck in.

Now loosen the rope nipper and let the wire rope close. Any slight inequality can be taken out by pounding the rope with a wooden mallet.

Next, shift the rope nippers, and repeat the operation at the other five places.

After the rope has run for a day, the locality of the splice can no longer be discovered. There are no ends turned under or sticking out, as in ordinary splices, and the rope is not increased in size, nor appreciably weakened in strength.

Notice to the Trade.

It has recently come to our notice that a Mr. James Richmond, of Lockport, N. Y., claims certain patent rights in connection with the rubber filling of wheels, the transmission of power in different directions, etc.

We hereby give notice that we agree to protect all of our customers against any claim Mr. Richmond may make for royalty, and warn them not to pay it.

The filling of the wheel, whether by rubber, leather or gutta percha was patented as long ago as 1855 by F. Hirn in all the States of the continent and England, and has already expired through lapse of time.

Previous to 1867, over one thousand Rope Transmissions had already been put up in France, Germany and Switzerland, embracing every variety of arrangement now claimed by Richmond as his inventions !

On page 174 of a German Text-book (The "Constructeur" by Reuleaux), edition of 1861, will be found full illustrations of rubber fillings, wheel sections, etc., identical with those described in the American Patent No. 61,554 of Jan. 29th, 1867, in which Mr. Richmond bought a sixteenth interest for twenty dollars, and constituting the sole foundation upon which he proposes to collect a royalty from our customers.

In the illustration of this same patent a chain of endless buckets is shown as operated by two endless wire ropes. This identical arrangement was used already as far back as 1859 in dredging out the bridge foundations at Kehl over the Rhine, and is described in "Erbkam's Zeitschrift für Bauwesen" for Jan. 1860. This dredge never was successful, because the motion is slow, showing that the supposed inventor never understood the principle of Rope Transmission, which is *speed*.

In the report of the U. S. commissioners to the Paris exhibition, 1867, vol. III, page 128 to 134, will be found a full description of "Hirn's" Wire Rope Transmission as there exhibited.

Again, during the years 1862 to '66, the magnificent Wire Rope transmission at Schaffhausen on the Rhine was erected. Here 800 horsepower are distributed for a distance of two miles among fifty different manufactories, located in every imaginable position and embracing all arrangements of changing direction which Mr. Richmond thinks he has newly devised. A description of these works, with drawings and plans complete, was published in 1866 by J. H. Kronauer, of Winterthur, Switzerland.

We first introduced the system of Rope Transmission into this country in 1867, solely with a view to benefit American manufacturers, and have carefully abstained from hampering the thing with patents, in order to extend its use as much as possible and make it free to all. Among others so benefited was Mr. Richmond, whom we furnished with a set of wheels and rope for an 800 foot Transmission, about a year after the first issue of this pamphlet.

Finding that it proved perfectly successful, he at once turns around and attempts not only to appropriate the labors of others to himself, but even to deprive them of the fruits of their own labor—designs that could only be accomplished through the want of information among the examiners at the Patent Office.

JOHN A. ROEBLING'S SONS.

Plate V.

ROEBLING'S WIRE ROPE FASTENINGS.

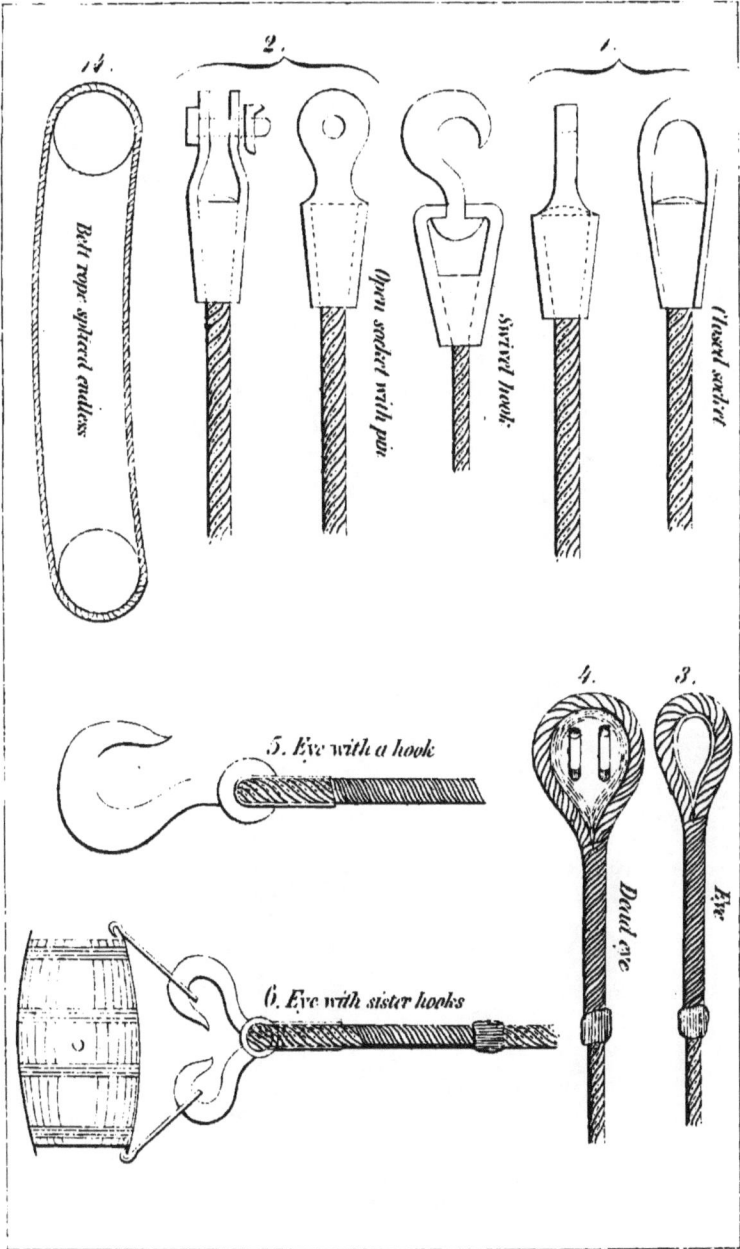

7. Belt rope spliced endless

2. Open socket with pin

Swivel hook

1. Closed socket

5. Eye with a hook

6. Eye with sister hooks

4. Dead eye

3. Eye

Plate VI
ROEBLING'S WIRE ROPE FASTENINGS.

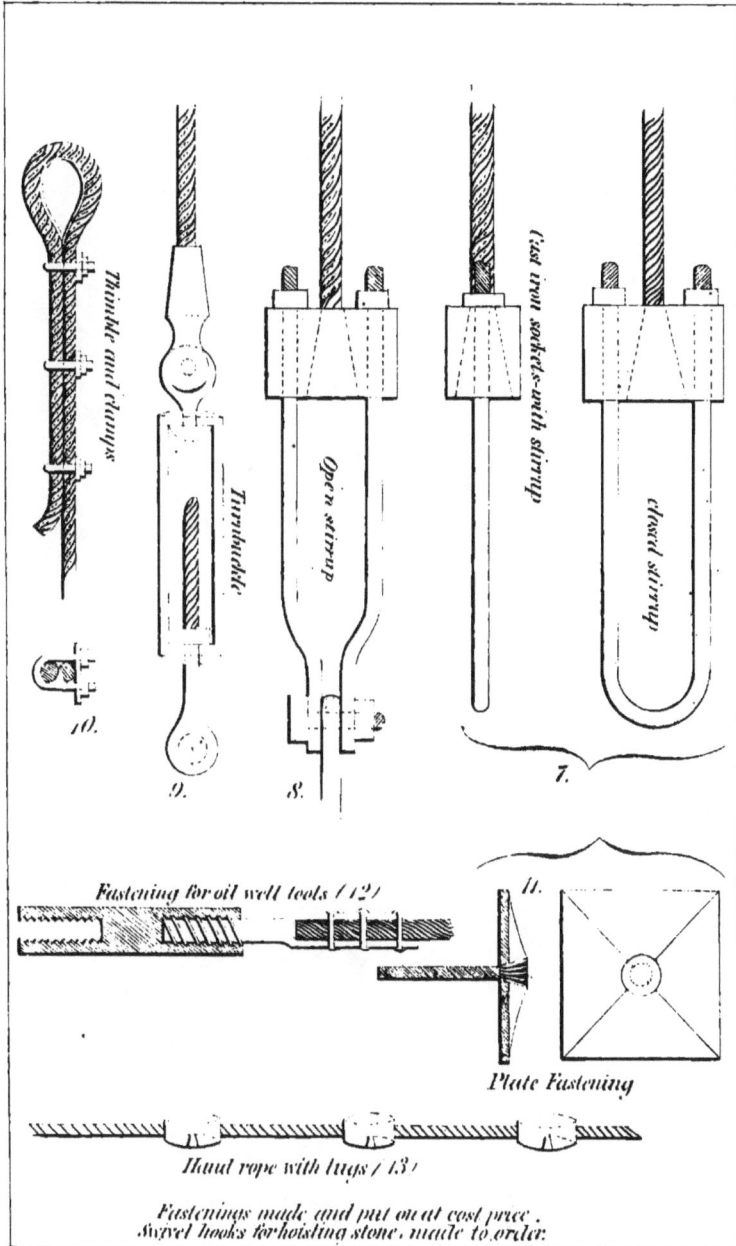

Thimble and clamps

10.

Turnbuckle

9.

Open stirrup

8.

Cast iron sockets with stirrup

closed stirrup

7.

Fastening for oil well tools (12)

11.

Plate Fastening

Hand rope with lugs (13)

Fastenings made and put on at cost price.
Swivel hooks for hoisting stone, made to order.

JOHN A. ROEBLING'S SONS.
MANUFACTURERS OF WIRE ROPE.
TRENTON, N.J.
STORE AND BRANCH OFFICE, 117 LIBERTY STREET, NEW YORK.

			IRON.		HOISTING ROPE, 19 WIRES TO THE STRAND.				STEEL.				
Trade Number.	Circumference in Inches.	Diameter.	Breaking Strain in Tons of 2000 Lbs.	Proper Working Load in Tons of 2000 Lbs.	Circumference of Hemp Rope of Equal Strength.	Min. Size of Drum or Sheave in Feet.	Price per Foot in Cents	Breaking Strain in Tons of 2000 Lbs.	Circumference of Hemp Rope of Equal Strength.	Proper Working Load in Tons of 2000 Lbs.	Min. Size of Drum or Sheave in Feet.	Price per Foot, in Cents.
1	6¾	2¼	74	15	15½	8	132	107		22	9	164
2	6	2	65	13	14½	7	115	97		20	8	144
3	5½	1¾	54	11	13	6½	100	78	15¾	18	7½	124
4	5	1⅝	44	9	12	5	86	64	14½	13	6	106
5	4⅜	1½	35	7	10¾	4½	71	53	13	11	5½	90
6	4	1¼	27	5½	9½	4	58	39	12½	9	5	74
7	3½	1⅛	20	4	8	3½	45	30	10	6	4½	57
8	3⅛	1	16	3	7	3	37	24	9¼	5	4	46
9	2¾	⅞	11½	2½	6	2¾	31	20	8¼	4	3¾	38
10	2¼	¾	8.64	2	5	2½	28	13	6½	3	3½	34
10¼	2	⅝	5.13	1¼	4½	2	26	7	5	2	3	33
10½	1⅝	9/16	4.27	¾	4	1¾	25	5	4¼	1¼	2¾	32
10¾	1½	½	3.48	½	3¾	1½	24					

Tiller Rope, ⅝ in. Diameter, 28c per Foot.

"   " ½ "   " 23c " "

All Kinds of Shackles, Sockets, Swivel Hooks, and Fastenings, put on, Splices made for Belt-Ropes.

IRON. *Rope with 7 Wires to the Strand.* *STEEL.*

Trade Number.	Circumference.	Diameter.	Circumference of Hemp Rope of Equal Strength.	Ultimate Strength in Tons of 2000 Lbs.	Proper Load in Tons of 2000 Lbs.	Price per Foot, in Cents.	Circumference of Hemp Rope of Equal Strength.	Ultimate Strength in Tons of 2000 Lbs.	Proper Load in Tons of 2000 Lbs.	Price per Foot, in Cents.	Trade Number.
11	4⅝	1½	10¾	36	9	60	13	50	12½	74	11
12	4¼	1⅜	10	30	7½	52	12	43	10	64	12
13	3¾	1¼	9½	25	6¼	45	10¾	36	9	55	13
14	3⅜	1⅛	8¼	20	5	39	9	29	7	47	14
15	3	1	7¼	16	4	32	8	23	6	40	15
16	2⅝	⅞	6¼	12.3	3	25	7½	18	5	32	16
17	2⅜	¾	5½	8.8	2½	20	6½	13	3¼	24	17
18	2⅛	11/16	5	7.6	2	17	5¾	11	2¾	20	18
19	1⅞	⅝	4¾	5.8	1½	14	5	8½	2¼	17	19
20	1⅝	½	4	4.1	1	12	4¾	6	1½	15	20
21	1⅜	7/16	3¼	2.83	¾	10					
22	1¼	⅜	2¾	2.13	½	9					
23	1⅛	5/16	2½	1.65		8					
24	1	9/32	2¼	1.38		7					
25	⅞	¼	2	1.03		6½					
26	¾	7/32	1¾	.81		6					
27	⅝	3/16	1½	.56		5½					

COPPER LIGHTNING RODS of all Varieties,

MADE TO ORDER.

We keep on hand a full assortment of Rubber Lined, Cast Iron Transmission Wheels, for transferring from 5 to 300 horse power any distance from 100 feet to 2 miles.

Send for pamphlet on "Transmission of Power."

STEEL CABLES FOR SUSPENSION BRIDGES.

Diameter in Inches.	Ultimate St'th in Tons of 2000 Lbs.	Weight per Foot.	Price per Foot in Cents.
2⅝	200	15	
2½	160	11	
2⅜	120	8.5	
2¼	107	7.4	
2	96	6.5	
1⅞	88	6	
1¾	75	5¼	
1⅝	61	4¼	
1½	50	3½	

IRON, COPPER AND TINNED SASH CORDS.

Trade No.	Diameter.	Iron.	Copper.	Tinned.
		Price per Foot.		
25	¼	6½	13	7
26	7/32	6	11	6½
27	3/16	5½	9	6
27½	5/32	4	6	5
28	⅛	3	4½	4
29	1/16	2	3½	3

Galvanized Charcoal Wire Rope
FOR SHIPS' RIGGING.

7 WIRES TO THE STRAND.					12 WIRES TO THE STRAND.				
Circumference.	Weight per Fathom.	Circumference of Hemp Rope of Equal Strength.	Breaking Strain in Tons of 2000 Lbs.	Price Per Pound.	Weight per Fathom.	Circumference of Hemp Rope of Equal Strength.	Breaking Strain in Tons of 2000 Lbs.	Price per Pound.	Circumference.
6in.	30℔	12in.	50	14	30℔	12in.	50	14½	6in.
5½	26	11	43	"	26	11	43	"	5½
5¼	24	10½	40	"	24	10½	40	"	5¼
5	22	10	35	"	22	10	35	"	5
4¾	20	9½	33	"	20	9½	33	"	4¾
4½	18	9	30	"	18	9	30	"	4½
4¼	16	8½	26	"	16	8½	26	"	4¼
4	14	8	23	15	14	8	23	15½	4
3¾	12	7½	20	"	12	7½	20	"	3¾
3½	10	7	16	"	10	7	16	"	3½
3¼	8½	6½	14	15½	8½	6½	14	16	3¼
3	7½	6	12	"	7½	6	12	"	3
2¾	6½	5½	10	15¾					
2½	5½	5	9	"					
2¼	4½	4½	8	16¼	SEIZING STUFF,				
2	3½	4	7	16½	ALL SIZES,				
1¾	2¾	3½	5	17¼	SIGNAL STRAND FOR R. R. USE,				
1½	2¼	3	3½	19¼	GALV. FENCING STRAND.				
1¼	1¾	2½	2½	21					
1	1¼	2	2	23					

NOTES ON RIGGING.

Galvanized Wire Rope for shrouds and stays is now universally superceding hemp rope for the following reasons: it is much cheaper; more durable; and will not stretch permanently under great strains, as is the case with hemp rigging, thus saving much labor in setting up; and it is fully as elastic as hemp rope of equivalent size. The great economy of using wire in place of hemp rigging is the large reduction in size and weight. The bulk of wire rigging is only one sixth that of hemp, while the weight is only one-half. The advantages of lightness are apparent to every seaman; it offers less resistance to the wind, and the removal of several tons of weight from the height occupied by the standing rigging, increases both the steadiness and stability of the ship.

All vessels in the U. S. Navy are now rigged with Roebling's Wire Rope exclusively, it having proved the best in the test made by the Government at the Washington Navy Yard.

Notes on the uses of Wire Rope.

Two kinds of wire rope are manufactured. The most pliable variety contains 19 wires in the strand and is generally used for hoisting and running rope. The ropes with 12 wires and 7 wires in the strand are stiffer, and are better adapted for standing rope, guys and rigging. Orders should state the use of the rope and advice will be given. Ropes are made up to 3 inches in diam., both of iron and steel, upon special application.

For safe working load allow one-fifth to one-seventh of the ultimate strength, according to speed, so as to get good wear from the rope When substituting wire rope for hemp rope, it is good economy to allow for the former the same weight per foot which experience has approved for the latter.

Wire rope is as pliable as new hemp rope of the same strength ; the former will therefore run over the same sized sheaves and pullies as the latter. But the greater the diameter of the sheaves, pulleys or drums, the longer wire rope will last. In the construction of machinery for wire rope it will be found good economy to make the drums and sheaves as large as possible. The minimum size of drum is given in a column in the table.

Experience has demonstrated that the wear increases with the speed. It is therefore better to increase the load than the speed.

Wire rope is manufactured either with a wire or a hemp centre. The latter is more pliable than the former and will wear better where there is short bending. Orders should specify what kind of centre is wanted.

Wire rope must not be coiled or uncoiled like hemp rope. When mounted on a reel, the latter should be mounted on a spindle or flat turn-table to pay off the rope. When forwarded in a small coil without reel, roll it over the ground like a wheel, and run off the rope in that way. All untwisting or kinking must be avoided.

To preserve wire rope, apply raw linseed oil with a piece of sheepskin, wool inside ; or mix the oil with equal parts of Spanish brown or lamp-black.

To preserve wire rope under water or under ground, take mineral or vegetable tar, add 1 bushel of fresh slacked lime to 1 barrel of tar, which will neutralize the acid, and boil it well, then saturate the rope with the hot tar. To give the mixture body, add some sawdust.

In no case should *galvanized rope* be used for running rope. One day's use scrapes off the coating of zinc, and rusting proceeds with twice the rapidity.

The grooves of cast iron pulleys and sheaves should be filled with well seasoned blocks of hard wood set on end, to be renewed when worn out. This end wood will save wear and increase adhesion. The smaller pulleys or rollers which support the ropes on inclined planes should be constructed on the same plan. When large sheaves run with very great velocity, the grooves should be lined with leather, set on end, or with india rubber. This is done in the case of all sheaves used in the *transmission of power* between distant points by means of ropes, which frequently run at the rate of 4000 feet per minute. Full information will be given, on the size of rope and the size and speed of sheaves to be used for transmitting power.

Steel ropes are to a certain extent taking the place of iron ropes, where it is a special object to combine lightness with strength.

But in substituting a steel rope for an iron running rope, the object in view should be to gain an increased wear from the rope rather than to reduce the size.

Send for Pamphlet on "Transmission of Power by Wire Rope."